Why Science Matters

Finding and Using Oil

John Coad

www.heinemann.co.uk/library
Visit our website to find out more information about Heinemann Library books.

To order:

☎ Phone 44 (0) 1865 888066

🖹 Send a fax to 44 (0) 1865 314091

💻 Visit the Heinemann Bookshop at www.heinemann.co.uk/library to browse our catalogue and order online.

First published in Great Britain by Heinemann Library, Jordan Hill, Oxford OX2 8EJ, part of Pearson Education Limited, a company incorporated in England and Wales having its registered office at Edinburgh Gate, Harlow, Essex, CM20 2JE – Registered company number: 00872828

© Pearson Education Limited 2009
The moral right of the proprietor has been asserted.

Editorial: Andrew Farrow, Megan Cotugno, and
           Harriet Milles
Design: Steven Mead and Q2A Creative Soultions
Illustrations: Gordon Hurden
Picture research: Ruth Blair
Production: Alison Parsons

Originated by Modern Age
Printed and bound in China by Leo Paper Products

ISBN 978 0 4310 4053 0
13 12 11 10 09
10 9 8 7 6 5 4 3 2 1

British Library Cataloguing-in-Publication data
Coad, John
  Finding and using oil. - (Why science matters)
  1. Petroleum - Juvenile literature 2. Petroleum -
  Prospecting - Juvenile literature
  I. Title
  553.2'82
A full catalogue record for this book is available
from the British Library

Acknowledgements
The publisher would like to thank the following for permission to reproduce photographs:
© Alamy **pp4** (M Stock), **25** (G P Bowater), **39** (Butch Martin); © Alamy/A Room With Views **p13**; © Corbis/Bettmann **pp5**, **17**; © Corbis/Brooks Kraft **p46**; © Corbis/Jean-Paul Pelissier/Reuters **p42**; © Corbis/Pat J. Groves/Ecoscene **p32**; © Corbis/Richard Melloul/Sygma **p19**; © Corbis/Reuters **p16**; ©Corbis/Robert van der Hilst **p27**; ©Getty Images/AFP Photo/Adrian Dennis **p45**; ©Imagestate/Robert Llewellyn **p15**; PA Photos **pp31** (AP/Duane Burleson), **8** (AP/Gustavo Ferrari), **43** (AP/Michael S. Green); **29** Pearson Education Ltd/Tristian Leverett; ©Photodisc/StockTrek **p47**; ©Photolibrary/Alaskastock/Mark Newman **p37**; ©Science Photo Library/Maximilian Stock Ltd **p11**. Background images supplied by ©istockphoto.

Cover photograph of an oil rig reproduced with permission of © Getty Images/Taxi and ©istockphoto.

The publishers would like to thank Andrew Solway for his invaluable assistance in the preparation of this book.

Contents

Some words are printed in bold, **like this**. You can find out what they mean in the glossary.

The quest for oil

Black gold

People have searched the world for it. They have fought and died over it. Some have got rich from it. We worry when its price goes up. We want more of it but know we shouldn't use too much. We're talking about a dark, sticky, smelly liquid called **crude oil** or **petroleum**. It is so valuable that it is often referred to as "black gold".

We depend on it

Our world runs on oil. Petrol for cars, diesel for lorries, and kerosene for planes all come from oil. We generate electricity by burning oil. Many of the things we take for granted are obtained from the chemicals in oil. These include cosmetics, paints, **synthetic** fibres, dyes, **plastics**, detergents, and even medicines. Without oil, our world would be a very different place.

How many things in this picture have been made from oil? The car's fuel, its paintwork, the seat materials, the sponge, the detergent, the plastic bucket, and the road surface all came from oil. Even the synthetic fibres and dyes in the man's clothes are made from chemicals in oil.

It has been around for a long, long time!

Around 3,000 years ago in ancient Egypt, oil was used to grease the axles of the pharaohs' chariots. Egyptians also used pitch, a thick form of oil, as a coating to help preserve mummies. Other ancient people used pitch to make their ships watertight.

Historians say the Chinese used oil around 300 BCE. They collected the oil with bamboo tubes and used it for lighting and cooking.

The start of the oil industry

Ignacy Lukasiewicz (1822–82) was born in Poland. His parents were poor, so he had to go to work at the age of 14. He worked as an assistant to a pharmacist (chemist). After a time, the pharmacist insisted that Ignacy should go to university in Krakow. After years of study he finally graduated in pharmacy.

At this time, the main form of lighting was oil lamps. The best lamp oil was whale oil, but it was expensive. Ignacy became interested in using the oil that seeped out of the ground as a cheap alternative. In 1853, he **distilled** clear kerosene from crude oil. Later that year he loaned one of his kerosene lamps to the local hospital for an emergency surgical operation. The following year he opened the world's first oil "mine". More oil wells were opened and in 1856 the first oil refinery was built. Ignacy Lukasiewicz became a wealthy man but used his wealth generously to help his community.

A 19th century oil well drilled in the United States. There was great pressure underground and the oil would shoot out in what was known as a "gusher". Technology at the time could not prevent these blowouts, which were very dangerous and wasteful.

Where did oil come from?

Oil is a **fossil fuel**. It is made from the fossilized remains of living things. Between 500 and 150 million years ago, conditions in parts of the Earth were favourable for oil formation. At this time, seas and swampy areas were rich in microscopic plants and animals. When these tiny living things died, they either settled straight to the seabed, or rivers and streams carried their remains, mixed with mud and sand, towards the oceans.

The dead creatures sank to the ocean floor and became covered with mud, sand, and other minerals. They did not decompose quickly as dead things do on the Earth's surface. This is because they were buried rapidly and no oxygen was present.

Over millions of years, the layers of sediment built up and the mixture was compressed and heated. At temperatures of more than 70°C (158°F), the dead creatures gradually decomposed into the simple chemicals we now find in oil.

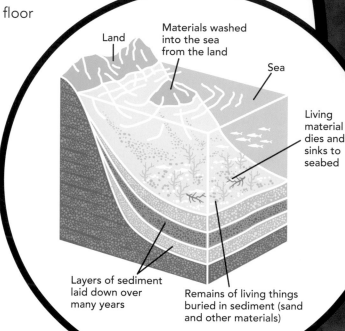

Land

Materials washed into the sea from the land

Sea

Living material dies and sinks to seabed

Layers of sediment laid down over many years

Remains of living things buried in sediment (sand and other materials)

CASE STUDY

A different view of oil formation

There is good evidence that oil came from living things but we cannot prove it. Although most scientists accept this theory, some do not. They believe the chemicals that make oil have always been in the Earth's crust. They believe that hydrogen and carbon were turned into the chemicals in oil under the high temperatures and pressures found deep inside the Earth during its formation. Over time they have gradually leaked up to the surface through cracks in rocks. We'll call this the inorganic theory – because it doesn't involve any living things.

Caps and traps

Sedimentary rock such as sandstone or limestone formed with the oil over millions of years. These rocks have pores – tiny holes and spaces between the particles that make up the rock. Oil soaked into the **porous rocks** leaving them saturated like a wet sponge. We call them **reservoir rocks**. Porous rocks also soak up water. Oil is less dense than water and will tend to rise to the surface. Oil will keep rising until it reaches a "cap rock". Cap rocks are **impermeable**. This means that liquid cannot soak through them. Clay is a good example of a cap rock.

As the Earth's crust (rocky surface layer) folded over, oil traps were created. This process took many, many thousands of years. Oil traps are places where reservoir rock is covered by cap rock and the oil cannot escape.

There must be an impermeable layer of rock capping the reservoir rock so the oil cannot flow away. The reservoir can be under great pressure in the Earth's crust. If we drill through the cap, oil could squirt out.

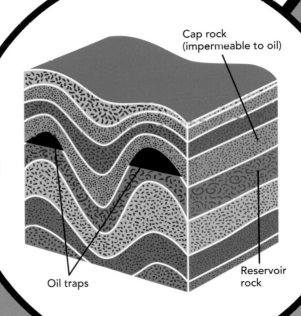

Cap rock
(impermeable to oil)

Oil traps

Reservoir rock

THE SCIENCE YOU LEARN: TYPES OF ROCK

There are three types of rock:
- *Igneous rock* is formed when molten material inside the Earth solidifies.
- *Sedimentary rock* is formed when deposits of sand, silt, and other materials are compressed together.
- *Metamorphic rock* is rock that has been changed by heat and pressure deep within the Earth's crust.

Black, brown, and honey-coloured

Crude oil is found in many countries, but its consistency and colour vary widely depending on the geological history of an area. In parts of the Far East, crude oil is waxy and black. It is similar to the oil found in central Africa because both were formed from similar sources. Crude oil from Western Australia can be a light, honey-coloured liquid. Many types of crude oil are found in the United States because there is great variety in the geological history of its different regions.

Kuwait's oil minister holding some light crude oil samples. The two samples are very different colours.

THE SCIENCE YOU LEARN:
IDEAS AND EVIDENCE

Our idea of how oil was formed is just a theory. Sometimes it is extremely difficult to prove scientific ideas. (See the Case Study on page 6.) We learn ideas and treat them as facts, even though we don't have the proof to support them. Think of police investigating a crime – they collect evidence, or clues, to help them form suspicions about who is guilty. However, clues are often not enough, and they have to find real proof. Scientists are the same. They find clues to help them develop theories, but it is often impossible to find real proof.

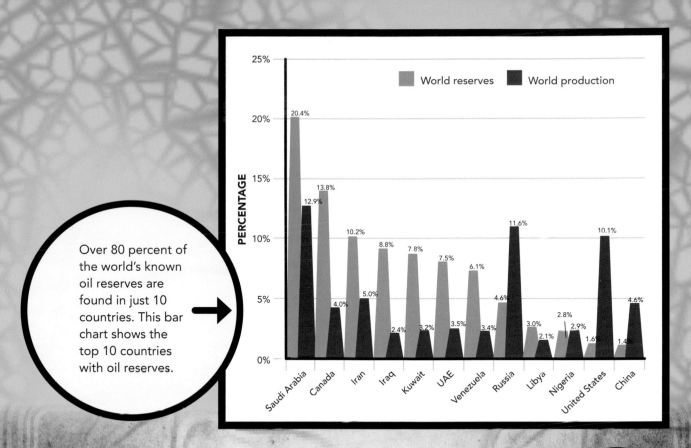

Over 80 percent of the world's known oil reserves are found in just 10 countries. This bar chart shows the top 10 countries with oil reserves.

Canada's oil reserves

Canada has the world's second-largest known oil reserves, but it is only the seventh-largest oil producer because its oil requires lots of expensive processing. In Alberta there are huge areas of sandstone in which oil is trapped, known as **tar sands**. If the sands are close to the surface they are dug out and processed to separate the tar from the sand, earth, and water. This process uses a lot of energy and water and has a huge impact on the environment. The mining operations have cleared thousands of hectares of trees and created pits 60 m (200 ft) deep. The region is dotted with large man-made lakes filled with mining waste. As demand for oil grows, it is likely that production will rise from about one million barrels per day now to four million per day by 2020.

Obtaining oil

Where is oil found?

We know that oil becomes trapped beneath caps, but how do these caps and traps form in the Earth's crust and what do they look like? Two of the most common traps are **anticlines** and **faults**.

Anticlines

When layers of rock fold to create a dome, the resulting structure is called an anticline. If porous reservoir rock is capped by an impermeable rock, oil may form at the crest. Oil floats on water, so the oil will rise to the top of the anticline above any water that is also soaked into the rock. Anticlines are the most common type of oil trap. The world's largest oil field is the Ghawar field in Saudi Arabia. It is formed by an anticline 240 km (150 miles) long. Anticlines are shown on page 7.

Faults

As the Earth's crust moves, layers of rock may break and slide past other layers. We call this a fault. If porous reservoir rock moves against an impermeable cap rock and the fault does not leak, the oil moves up and accumulates against the fault.

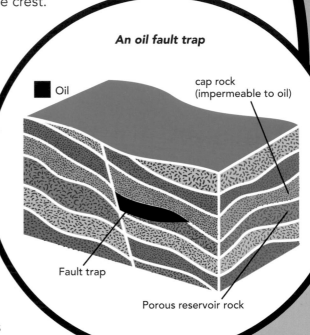

An oil fault trap

■ Oil

cap rock
(impermeable to oil)

Fault trap

Porous reservoir rock

IN THE NEWS: BARRELS OF OIL

Listen to the news and you will hear the price of crude oil quoted in US dollars per barrel. Many years ago, crude oil was transported in wooden whiskey barrels. This unit of measurement has remained unchanged despite the replacement of wooden barrels by larger steel barrels, and then by pipelines. One barrel of oil is equal to 159 litres (42 gallons). One tonne (1.1 tons) of oil is approximately equal to 7.3 barrels.

Prospecting for oil

When oil was first found, people knew where to dig because they could see the oil seeping from the ground. Today, oil companies use advanced technology to find new deposits. Scientists use **seismology** to learn about rock formations below the Earth's surface. A seismic survey uses a small explosion or a thumping or vibrating machine to send powerful sound vibrations into the ground. The vibrations reflect (bounce back) from boundaries between different kinds of rock underground. Sensors called geophones pick up the reflected sound waves, turn them into electronic signals, and feed them into a computer. The computer uses the information to produce a picture or **seismogram**. This shows a slice of the Earth from the surface downwards.

A computer-generated three-dimensional image of part of the Earth's crust. It shows several layers and where oil might be found.

Ian Ward (born 1972), geophysicist

Ian Ward works for a company that is involved in the exploration, production, transportation, and distribution of oil and gas. Although his job title is Principal Geophysicist, his current job is to interpret seismic data and create three-dimensional models of rock structures more than 5 km (3 miles) beneath the sea. He helps identify where to drill, finds out how much oil or gas is present, and works out what data to collect.

Widening the search

In 2007, Russian explorers reached the seabed at the North Pole in two mini-submarines. At 4,200 m (13,780 ft) below the surface, they planted the Russian flag and claimed the Arctic. Why was Russia so keen to own the frozen wastelands of the north?

Many countries with warmer climates are rapidly running out of oil. However, it is estimated that one-quarter of the world's undiscovered oil reserves lie under the Arctic Ocean. While Russia moves in from one side of the Arctic, the United States is prospecting (searching) north of Alaska. Environmentalists are worried about the impact of oil extraction on one of the last unspoiled regions of the world.

CUTTING EDGE: CSEM

Seismic mapping is very useful but cannot distinguish between rocks saturated (filled) with water and those containing oil. Now a new technique is being used for underwater exploration – Controlled Source Electromagnetic Imaging (CSEM).

In CSEM, low-frequency radio waves are transmitted into the seabed. The signals reflect back from underground rocks, as in seismic mapping. However, the reflected signals vary depending on how well the rocks conduct electricity. Oil deposits conduct electricity poorly, so they show up clearly in a CSEM image.

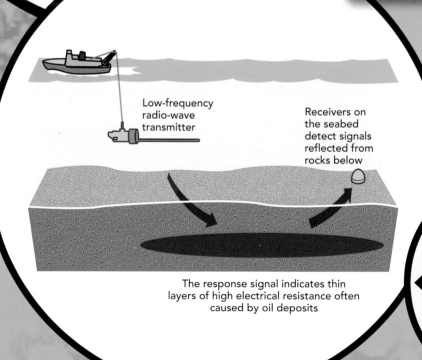

Low-frequency radio-wave transmitter

Receivers on the seabed detect signals reflected from rocks below

The response signal indicates thin layers of high electrical resistance often caused by oil deposits

Seismic studies are based on the acoustic properties of rock layers. CSEM detects fluids based on their electrical properties, not rock structures.

Locating an oil trap

Seismic surveys locate possible oil traps. Once a promising site has been selected, a drilling rig is set up. The crew move in and work starts.

In the first stage of drilling, the drill is lowered. The drill is attached to a drill pipe and it is rotated by a turntable. As the hole deepens, extra lengths of drill pipe are attached. Drilling can take weeks or months before the target is reached.

Directing the drill bit (the cutting tip of the drill) to a precise location – perhaps several kilometres away – requires advanced computer technology. A navigation device above the drill bit gives the exact position of the well. A motor in the drill pipe can adjust the direction of the drill.

Drilling fluid – mainly a mix of soil and water – is pumped down the drill pipe and into the hole. This washes the cuttings of rocks back to the surface and cools the drill bit. It also keeps a downward pressure on the well to prevent oil shooting uncontrollably to the surface.

Members of the drill crew attach another 9.5 m (30 ft) length of drill pipe. The pipe is 10–13 cm (4–5 inches) in diameter. The whole length of drill pipe is called the drill string. At the bottom end it attaches to the drill bit, and at the top it is turned by an engine.

CUTTING EDGE: DRILL TIPS

The most important part of a drill is the tip. This must cut through the hardest rock, deep in the Earth's crust. Drill bits range from about 90 cm (35 in) in diameter at the start of the hole to 20 cm (8 in) deeper down. They usually have cutting wheels tipped with either tungsten carbide or diamond.

Extracting the oil

Once the drill bit reaches the oil reservoir, oil may shoot to the surface. Pressure underground – perhaps from water rising up through the porous rock – forces oil up. This pressure doesn't last long and engineers soon have to install pumps to help bring up the oil. Alternatively they pump water or gas into the well to bring the oil to the surface. Even then a lot of oil is left behind. Sometimes another pipe is drilled nearby and steam pumped down. The steam turns to water and helps to push the oil up. The heat from the steam makes the oil less viscous (less thick). Even after all of these methods have been used, only about one-third of the oil in the reservoir is extracted.

Transporting oil

Each year about 3.5 billion tonnes (3.9 billion tons) of crude oil is produced worldwide. About half of this is exported from the Middle East, Africa, and Latin America to North America, Europe, and Southeast Asia. Most of this oil is transported by sea in giant tankers. It is estimated that there are over 3,500 oil tankers operating worldwide. The average voyage for a tanker lasts six weeks and involves at least one passage through an area in which many ships pass close together. In these places there is a high risk of an accident.

Transportation by pipeline is safer. However, leaks can happen as a result of carelessness or sabotage (deliberate damage). Pipelines have always been used to transport oil from the production fields to ports where it can be piped into tankers. Now some pipelines have been constructed to transport oil long distances, but it is not possible to move all oil in this way.

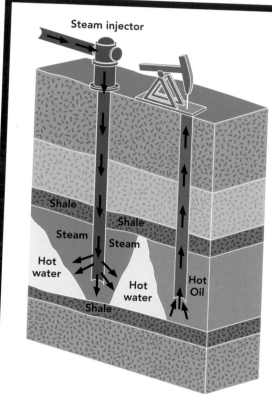

Steam injector

Shale
Shale
Steam
Steam
Hot water
Hot water
Hot Oil
Shale

This diagram shows the steam flood process for extracting oil. The heat from the steam helps to push it up the well.

The Knock Nevis supertanker is the world's largest ship. It is more than 450 m (1,476 ft) long and nearly 70 m (230 ft) wide. It is too large to navigate the English Channel, let alone the Panama and Suez canals. Today it is used as an offshore floating platform for the oil industry.

The longest pipeline

The Druzhba pipeline is the world's longest oil pipeline. It carries oil 4,000 km (2,485 miles) from south-east Russia to the Ukraine, Hungary, Poland, and Germany.

The pipeline was constructed in 1964. The name "Druzhba" means "friendship" and the pipeline supplied oil to friendly, energy-hungry western regions of the former Soviet Union. Today, it is the main method of moving Russian oil across Europe, transporting up to 1.4 million barrels per day.

Its length is soon to be exceeded by a new pipeline running from East Siberia to the Pacific Ocean. It will be 4,200 km (2,610 miles) long and will take Russian oil to China, Korea, and Japan.

Oil spills and pollution

Oil that is released into the environment causes pollution, especially in water. Oil tanker accidents sometimes cause large spills, but smaller amounts are released when ships wash out their oil tanks and dump oil in the sea. The biggest source of oil pollution is waste from cities and factories. This drains into rivers and eventually reaches the ocean.

Spilt oil floats on water. This is easier to clean up than if it sinks and forms a layer on the seabed. Even so, surface oil harms sea birds. In rough seas oil eventually breaks into small droplets and sinks, where it can harm other sea creatures. Following bad spillages fish have been caught with severe skin ulcers, and some have even tasted of oil.

THE SCIENCE YOU LEARN: OILED SEABIRDS

Contact with crude oil causes sea birds severe problems:
- Feathers collapse and mat together. This can affect the ability of the bird to fly.
- Feathers no longer keep the birds warm.
- Skin may become severely irritated.
- Birds often eat the oil when they try to preen (clean) themselves. Consuming oil poisons the birds.

THE WORLD'S BIGGEST OIL TANKER SPILLS

Tonnes of oil (approx)	Spill date	Tanker	Location
220,000	1978	Amoco Cadiz	Brittany, France
160,000	1979	Atlantic Empress	Tobago, Caribbean Sea
120,000	1967	Torrey Canyon	Cornwall, UK
85,000	1993	Braer	Shetland Isles, UK
72,000	1996	Sea Empress	Milford Haven, UK
39,000	1989	Exxon Valdez	Alaska, USA

One of the world's worst oil disasters

In 1979, the Mexican oil company PEMEX was drilling beneath the sea in the Gulf of Mexico. The drill had reached 3,627 m (11,900 ft) below the seabed when disaster struck. The IXTOC 1 well suffered a blowout – the pressure in the well was greater than the pressure in the drill pipe and the safety systems failed. Gas and oil shot to the surface where they caught fire and engulfed the drill platform. The oil rig collapsed, sank to the seabed, and littered the area with broken debris and 3,000 m (9,840 ft) of pipe.

The well initially released 4,000 tonnes (4,400 tons) of oil per day. Efforts to plug the well reduced this to 1,400 tonnes (1,540 tons) per day. Two new wells were drilled close by to relieve the underground pressure but it took nine months to completely stop the flow. At its worst, an oil slick 180 km (112 miles) long and 80 km (50 miles) wide covered the ocean. Around 500 air missions were flown to spray dispersant chemicals on the slick, but winds carried the oil to the coast of the United States. Texas suffered the greatest damage. This accident was the second biggest ever single oil spill and it released around 480,000 tonnes (529,100 tons) of oil.

IXTOC 1 exploratory well blew out on 3 June 1979 in the Bay of Campeche of Cuidad del Carmen, Mexico.

Cleaning it up

Here are some of the methods for cleaning oil spills:

Booms

It is easier to clean up oil if it is all in one place. Containment booms act like a floating fence to keep the oil from spreading.

Skimmers

Skimmers suck up oil like a vacuum cleaner. Skimmers get clogged easily and do not work well when the water is rough.

Adsorbents

Oil will coat some materials – this is adsorption. Straw is a very simple adsorbent. However, once the material is coated in oil, it may sink. The material would then stay on the seabed where it could harm wildlife.

Absorbents

These are materials that soak up liquids. They are used to soak up oil from the water's surface or from onshore oil-coated rocks.

Chemicals

Detergents break oil into lots of small droplets. The oil droplets disperse or mix with the water. However, detergents can cause more pollution and harm wildlife, and the dispersed oil remains in the water.

Burning

Burning can remove up to 98 percent of an oil spill. The spill must be a minimum of 3 mm (0.12 in) thick and relatively fresh for this method to work. Burning also releases smoke, causing air pollution. On land, burning can make land barren for years.

Biological

Some bacteria naturally break down oil. However, it takes years for oil to be removed in this way. Adding a fertilizer or microorganisms to the water can speed up the process. The fertilizer gives the bacteria the nutrients they need to grow and reproduce. Adding microorganisms increases the population that is available to degrade the oil.

INVESTIGATION:
HOW EASY IS IT TO CLEAN UP AN OIL SPILL?

Float some cooking oil on water and add different materials to see which soaks up the oil best. Before you carry out the experiment, you will have to decide how to measure the amount of oil removed, what equipment to use, and how to carry out the experiment. You will need to think about the variables in the experiment. What would you keep constant? The amount of oil? The amount of water?

Researchers at a university in Egypt carried out such an investigation. They wanted to see if there were any cheap, easily available, biodegradable, **recyclable**, natural materials that could be used to clean up a spill. They used a plant called *Cynanchum acutum*, which grows on sandy shorelines in northern Africa. The seeds of this plant are covered in white silky hairs (WSH).

The researchers used a standard volume of salty water with a fixed amount of Egyptian crude oil. The WSH were used to absorb the oil and were then removed from the surface. The water was shaken with a solvent to remove any remaining oil. The colour of the solvent and oil mixture was measured by a machine. In this way the scientists could find out exactly how much oil was left. They found that the WSH absorbed three times their own mass of oil. By doing this investigation, they showed that the plant material could be a cheap, easy, and natural way to clean up oil spills.

A member of the emergency services sprays rocky coastline with detergent following an oil spill.

The chemistry of oil

Hydrocarbons

Oil is a mixture of many substances. These are mainly compounds (see box) made from just two elements, carbon and hydrogen. We call these compounds **hydrocarbons**.

Carbon is an unusual element because its atoms can link together to make chains and rings. This means that there are millions of possible hydrocarbons with different chain lengths and shapes. One important rule is that carbon atoms always make four bonds (links). The simplest hydrocarbon is therefore one carbon atom linked to four hydrogen atoms. It is called methane and has the chemical formula CH_4. Methane is found in natural gas, which we use for our heating and cooking. There is a whole series of compounds like methane. We call them **alkanes** and each of their names ends in -ane.

THE SCIENCE YOU LEARN: ELEMENTS AND COMPOUNDS

All the substances we know are made from chemical elements. Elements are the simplest substances that can exist. The atoms of each element are unique. There are over 100 different elements, each with a symbol to represent its atoms. Elements join together to make new substances called compounds, which have completely new characteristics. The atoms in a compound are held together by bonds. These bonds are formed when atoms share or transfer electrons.

The alkanes

The table below gives information on the first four members of the alkane series. Can you see the pattern in the formulas? It is: double the number of carbon atoms and add two to work out the number of hydrogen atoms. The first four hydrocarbons are all gases – the liquids in crude oil are larger alkane molecules. For more information on larger alkanes see "Find out more" on page 48.

Real molecules are three-dimensional. The pictures show the three-dimensional shapes of the first four alkanes. They are difficult to draw like this so chemists usually flatten them out and draw bonds at right angles.

The first four alkanes

Methane CH_4		
Ethane C_2H_6		
Propane C_3H_8		
Butane C_4H_{10}	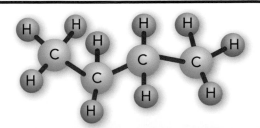	

Valuable, but useless as it is

If you ignite crude oil it will burn with a very smoky flame and leave behind a sticky black residue. Crude oil straight out of the ground is useless as a fuel. It has to be refined. In a refinery the substances in oil are separated into useful products.

To separate it, oil is first heated to about 350°C (660°F). Most of the liquids in the oil turn to vapour. The mixture is piped into the base of a huge tower called a fractionating column. Anything that is still liquid falls to the very bottom of the column. The hot mixture of gases rises and begins to cool. As it cools, the substances with the highest boiling points condense (turn back into liquid). Compounds with lower boiling points have to rise higher and cool further before they condense. Separating a mixture of liquids in this way is called **fractional distillation**.

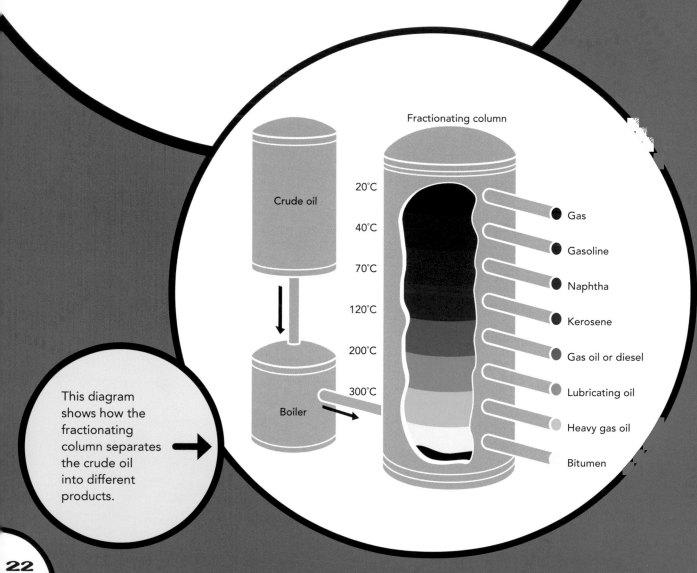

This diagram shows how the fractionating column separates the crude oil into different products.

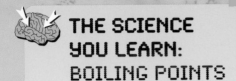

THE SCIENCE YOU LEARN: BOILING POINTS

When a liquid is heated its temperature rises until it reaches its boiling point. Then bubbles of vapour form in the liquid and the temperature will not go higher. Boiling point is also the temperature at which vapour condenses. Different liquids boil at different temperatures. For example, water boils at 100°C (212°F) but alcohol boils at 78°C (172°F).

Inside the column

There are metal trays, like shelves, inside the fractionating column. Holes in the trays allow vapour to pass upwards. Liquids with similar boiling points condense on the trays. Each tray is fitted with devices called "bubble caps" to prevent liquid falling back down the hole and to ensure the vapour and liquid mix thoroughly.

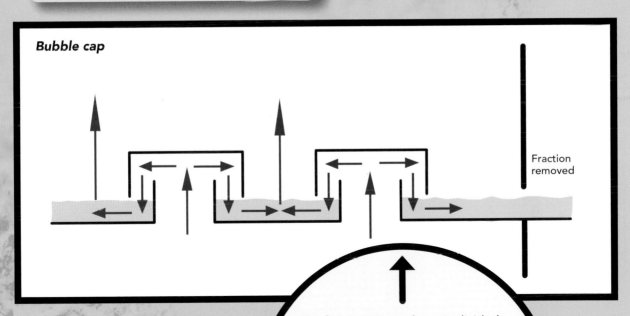

Bubble cap

Fraction removed

The fractionating column is divided by several trays. Hot vapour rises and passes through bubble caps on the trays. The bubble caps force the vapour to bubble through liquid on a tray. Substances that condense at that particular temperature join the liquid while those with lower boiling points rise to the next tray.

The fractions in oil

There are thousands of different compounds in crude oil. It is impossible to separate oil into individual, pure substances. Fortunately there is an important scientific principle that helps:

Similar sized hydrocarbon molecules have similar properties.

If we collect substances with boiling points that are similar, we will still have a mixture but the compounds in it will behave in a similar way. We call these mixtures "fractions". The table below summarizes the fractions in crude oil.

Name of fraction	Number of carbon atoms	Boiling-point range	Colour	Viscosity	Uses
Petroleum gases	1–4	Less than 20°C (68°F)	Very light in colour	Gas	Bottled gases
Petrol (gasoline)	5–10	20–70°C (68–158°F)	Increasingly dark in colour	Very runny	Fuel for cars
Naphtha	8–13	70–190°C (158–374°F)		Increasingly thick	Valuable source of chemicals
Kerosene (paraffin)	10–16	120–240°C (248–464°F)			Aeroplane fuel
Diesel	15–25	220–260°C (428–500°F)			For trains and lorries
Lubricating oil and fuel oil	20–70	250–350°C (482–662°F)			In engines and machinery Fuel for ships and central heating
Bitumen	More than 70	Does not vaporize easily	Black	Thick, sticky solid	Road surfaces, roofing

THE SCIENCE YOU LEARN: UNDERSTANDING VISCOSITY

Some liquids are thicker or more syrupy (viscous) than others. Why do they vary in this way? Think of hydrocarbon molecules in oil. If they are made of long chains of carbon atoms then you might expect them to tangle up. This is rather like a plate of long pieces of spaghetti. They are harder to separate. With hydrocarbon molecules there is a simple link – the bigger the molecules, the higher the boiling point, and the more viscous the liquid.

CASE STUDY

The Jamnagar oil refinery

Sitting on the edge of India's western coast, 800 km (500 miles) north of Mumbai, is the Jamnagar oil refinery. It is owned by Reliance Industries, India's largest private company. Building work started in 1996 with 100,000 workers toiling around the clock. The first phase was to build a refinery to process 660,000 barrels of oil per day. The second phase of construction was to double this capacity to 1.2 million barrels per day, making Jamnagar the world's largest refinery.

Forty percent of the petrol (gasoline) produced at Jamnagar will be shipped 15,000 km (9,320 miles) to the United States.

Fuels

Burning fuels

Fuels are useful and valuable because they burn. For fuels to burn they need oxygen. Combustion (burning) is a chemical reaction that releases energy. There are risks involved in using fuels. A spark in the wrong place at the wrong time can lead to a disaster. Hundreds of people die each year from such accidents in the home or workplace.

THE SCIENCE YOU LEARN: THE FIRE TRIANGLE

The fire triangle (see left) is a simple model from the science of firefighting. Three things are needed to start a fire: fuel, oxygen, and heat. These three conditions are represented in the fire triangle – take away one and a fire cannot happen.

IN THE NEWS: TURN OFF YOUR MOBILE PHONE!

Flames shot up around a 21-year-old college student whose mobile phone rang while he was filling his car with fuel. Firefighters said Matthew Erhorn was fortunate to receive only minor burns. "I'm very surprised," Matthew said.

He shouldn't have been surprised. Signs at the pumps at all petrol stations warn that mobile phones should be turned off. Firefighters believe the mobile phone ignited vapours coming from the car's fuel tank as it was being filled. It doesn't take much of a charge to ignite petrol vapours. Anything that can produce a spark could do it – such as a mobile phone.

The fires of Kuwait

In 1991, Iraqi forces retreating after their invasion of Kuwait set fire to oil wells across the country. Fires covered the desert landscape and black smoke filled the sky.

The world's most famous firefighter, Red Adair, was called in to tackle the problem. Red Adair had made his reputation by extinguishing a huge gas-field fire in the Sahara Desert that had been burning for six months. His team had also extinguished countless oil well fires around the world, including the world's worst offshore accident – the Piper Alpha disaster. In 1988, 167 people died when Piper Alpha, the largest oil rig in the North Sea, caught fire.

"Kuwait was easy," Red Adair confidently explained. "We put all the fires out with water, just went from one to the next." It wasn't that easy. To tackle a fire, everything needed to extinguish the flames and cap the well had to be in place. This was not easy in a desert area that had been through war. Huge volumes of seawater were pumped along oil pipes and Red Adair proceeded at an amazing pace. In just nine months, more than 600 fires were extinguished.

Red Adair used his understanding of the fire triangle to fight the oil fires in Kuwait. Water was used to cool the fires. The steam produced pushed oxygen away. Once the flames were out, the wells were capped to remove the fuel before the vapours could ignite again.

The chemistry of combustion

When a hydrocarbon is burned, the carbon gains oxygen to become carbon dioxide gas. The hydrogen gains oxygen to become water. We don't see the water because it is a vapour. This can be represented by the following equation:

Hydrocarbon + Oxygen ⟶ Carbon dioxide + Water

Every hydrocarbon will follow the same pattern:

Methane + Oxygen ⟶ Carbon dioxide + Water

Now look at what happens to the atoms in these compounds:

If we just mix methane and oxygen, nothing happens. Energy (a spark) is needed to begin breaking bonds. Once bonds in the reactants are broken, new ones can form, which releases more energy and keeps the process going.

From this we can write a chemical equation using symbols. The only other thing we need to know is that oxygen atoms join together in pairs. We write oxygen gas as O_2.

$$CH_4 + 2O_2 \longrightarrow CO_2 + 2H_2O$$

A similar equation can be written for any burning hydrocarbon.

THE SCIENCE YOU LEARN: CHEMICAL EQUATIONS

Chemists use chemical equations to represent a change that takes place. However, a proper equation tells us more than just what reacts and what is produced. When we write an equation properly, we must show the same number of atoms on each side of the equation. We call this a balanced equation. It tells us how much of each substance is involved. In the example above you can see that when methane burns it needs two molecules of oxygen and makes two molecules of water.

Where does the energy come from?

When we burn a fuel to cook, heat our home, or power our vehicles, where does the energy actually come from? To answer this we need to know more about the molecules that make up fuels.

Take the simplest hydrocarbon – methane (CH_4). One carbon atom has four bonds to hydrogen atoms. When we burn methane, the bonds have to be broken. We also have to break the bonds that hold oxygen atoms together. This takes energy. Carbon dioxide and water are formed. Making new bonds in these substances releases energy.

The bonds in carbon dioxide and water are strong. A lot of energy is given out when we make these products. The bonds in methane and oxygen are weak. It takes less energy to break the bonds in the reactants. In this case the balance between the making and breaking of bonds means that overall, energy is released.

Pulling magnets apart requires energy. When they attract each other they move and bang together, releasing energy. Chemical bonds behave a little like this.

THE SCIENCE YOU LEARN: EXOTHERMIC REACTIONS

Combustion is a chemical reaction that releases heat energy. This is an **exothermic** reaction, which means that heat is given out. Most chemical reactions are exothermic. Reactions that take in energy are endothermic.

Investigating fuel economy

In a car engine, the fuel burns and releases energy. How efficient is the engine? Does all the fuel burn? How much of the energy is converted into the movement of the car? How much is wasted?

Motorists have always been interested in the fuel efficiency of their vehicles. How would you measure the fuel economy of a car? At first the process seems really easy:

- Fill your tank up to the brim.
- Drive until the car's fuel tank is about one-quarter full and record the distance travelled.
- Fill up to the brim again, measuring exactly how much you put in.
- Divide the distance travelled by the volume of fuel used.

This gives the owner an idea of the car's fuel economy, but how accurate is it?

THE SCIENCE YOU LEARN: FAIR TESTING

When we compare the fuel economy of different cars, we want to know that tests have been fair. One variable (the type of car) is changed and another variable (the fuel economy) is measured. To make this fair we must keep all other variables the same. There are several factors that could affect fuel economy. These include:

- Road surface
- Tyre pressure
- The driver
- Hills
- Speed
- Wind
- Number of passengers

For the test to be fair, these must all be kept constant.

A vehicle performance technician tests the aerodynamic efficiency of a vehicle, using a device called a smoke wand.

Making a fair comparison

When manufacturers test new vehicles, the vehicle does not go on the road and no one measures the amount of fuel used! Instead, exhaust gases are collected, measured, and analysed. The tests are conducted in special laboratories and the cars rest their wheels on rolling roads. A rolling road is a set of rollers that are spun by the drive wheels. The driver has to drive along a virtual route determined by law. In Europe, the route is about 11 km (7 miles) long, takes 20 minutes to complete, and is divided into two parts called Urban and Extra-Urban.

At the end of the test, four bags of gas are taken for analysis. Two are filled with the exhaust produced by the car during the Urban and Extra-Urban periods of the test. The other two are filled with air taken from the lab during each period. The difference between the air in the laboratory and the exhaust emission is measured. A computer program takes the figures from the analysis, calculates how much fuel must have been used, and determines the fuel economy for the vehicle.

Green issues

Pollution

In a perfect world, burning a hydrocarbon fuel in a vehicle engine would only produce carbon dioxide and water. In practice this doesn't happen. Running a car engine produces a number of **pollutants**.

Some of the pollutants are **toxic**, but only once they reach a certain level. Government agencies monitor the air in our cities to measure the levels, but some evidence suggests that air pollution is still harming people.

Carbon monoxide (CO). This is produced whenever a hydrocarbon is burned in a limited supply of air, such as in a car engine. Carbon monoxide is extremely poisonous and has no colour or smell. It prevents blood from transporting oxygen around the body.

Nitrogen oxides (NOx). Nitrogen and oxygen make up 99 percent of our atmosphere. Normally they do not react together but a spark can make them combine. This happens in a car engine and forms either nitrogen oxide (NO) or nitrogen dioxide (NO_2) – we represent them in general by NOx. These oxides contribute to **acid rain**.

Particulates. These are microscopic soot particles produced by the combustion process. Experts do not know what levels are safe.

CUTTING EDGE: AIR POLLUTION AND ASTHMA

In 2002, a study in the United States showed that children who played team sports when ozone pollution levels were high were more likely to develop asthma. This supports an earlier study carried out in The Netherlands in 1999. There it was shown that breathing disorders in children worsened as air pollution increased. When particulates were in the air, children were most likely to suffer.

Sulphur dioxide (SO_2). Crude oil and the fractions obtained from it are not pure. One impurity in oil is sulphur. Burning sulphur produces sulphur dioxide. Sulphur dioxide is a major cause of acid rain, but only a small proportion of it comes from transport.

Benzene. An engine may not burn all the hydrocarbons in the fuel. Some might be emitted in the exhaust gases. Some unburnt hydrocarbons – particularly benzene – are known to cause cancer.

Ozone (O_3). In the Earth's upper atmosphere, the ozone layer protects us from harmful ultraviolet radiation. At ground level ozone is harmful to us. Ozone is not made directly in a car engine but is created by the action of sunlight on other pollutants.

Can air pollution kill?

Research carried out by the World Health Organization (WHO) indicates that more people die from the effects of pollution caused by vehicles' emissions than from road accidents. The study in France, Switzerland, and Austria estimated that 21,000 deaths per year were caused by long-term exposure to pollutants – more than twice the number killed in accidents. However, some scientists dispute the findings, suggesting there are other causes for the deaths.

Catalytic converters

In many countries, governments have passed laws that restrict the amounts of pollutants a vehicle can produce. Over the years, manufacturers have improved car engines and fuel systems to keep within these laws. One of these improvements has been the development of catalytic converters. The job of a catalytic converter is to turn harmful pollutants into less harmful gases before they leave the car's exhaust system.

Honeycomb structure coated with catalyst

Gases from engine:
Nitrogen oxides
Hydrocarbons
Carbon monoxide

Carbon dioxide
Water vapour
Nitrogen

A three-way catalytic converter.

Catalytic converters are surprisingly simple devices. They consist of a cylindrical body filled with a honeycomb structure made of a ceramic material. This has been coated with the **catalyst** – a mixture of platinum, palladium, and rhodium metals. The honeycomb structure creates a huge surface area (equivalent to three football pitches) inside a relatively small space.

Most modern cars are fitted with a three-way converter as part of their exhaust system. Firstly, the converter turns oxides of nitrogen back into harmless nitrogen and oxygen gases. Secondly, it oxidizes unburned hydrocarbons and carbon monoxide. This forms carbon dioxide and water. Thirdly, the converter has sensors to monitor the exhaust flow and to adjust fuel and air intake for the greatest fuel efficiency.

Our changing atmosphere

By burning huge amounts of fossil fuels (coal, gas, and oil) humans have added extra carbon dioxide to the atmosphere. Destroying vast areas of rainforest and burning trees has also released more carbon dioxide.

The greenhouse effect

The temperature inside a greenhouse is higher than outside. Energy from the Sun passes through the glass and warms up the inside. The glass prevents heat energy from escaping. It is trapped inside and the greenhouse stays warm. Carbon dioxide and other gases cause a similar effect around the Earth. Most scientists believe that increases in carbon dioxide levels are causing an enhanced greenhouse effect, leading to worrying changes in climate.

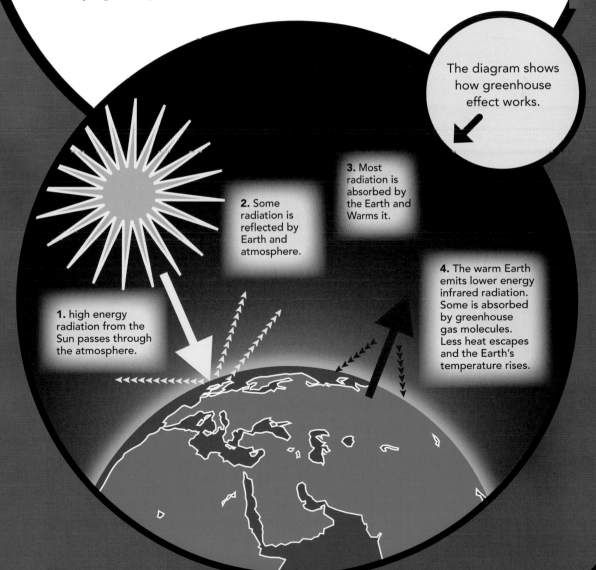

The diagram shows how greenhouse effect works.

2. Some radiation is reflected by Earth and atmosphere.

3. Most radiation is absorbed by the Earth and Warms it.

4. The warm Earth emits lower energy infrared radiation. Some is absorbed by greenhouse gas molecules. Less heat escapes and the Earth's temperature rises.

1. high energy radiation from the Sun passes through the atmosphere.

Carbon dioxide monitoring

The Mauna Loa Observatory (MLO) is located on the north side of
Mauna Loa Volcano, on the Big Island of Hawaii. MLO is a perfect place
for sampling the Earth's atmosphere. It is in a remote location in the
Pacific Ocean, 3,397 m (11,150 ft) above sea level, and far away from
major pollution sources. Being high on the volcano, the observatory
avoids pollutants that sometimes gather at sea level.

MLO began continuously monitoring and collecting data related to
climate change, the content of the atmosphere, and air quality in the
1950s. Today, the observatory is best known for its measurements
of rising carbon dioxide concentrations in the
atmosphere. This monitoring began in 1958
and has been recorded ever since. It forms
the longest available record of direct
measurements of carbon dioxide. The
results are best shown on a graph.

The graph shows how
carbon dioxide levels have
risen since the 1960s. Many
scientists believe that
human activity has been
the most significant cause
of this change.

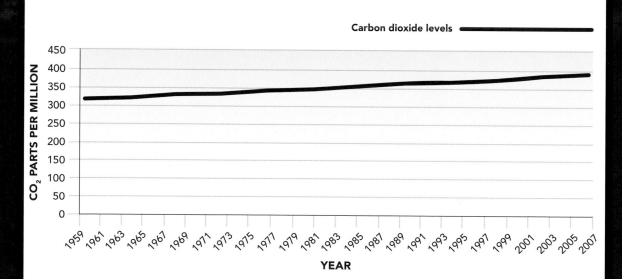

Atmospheric CO$_2$ each January at Mauna Loa Observatory

Carbon dioxide levels ▬▬▬▬▬

Climate change – it's happening

Al Gore (born 1948), the former Vice President of the United States, released a documentary film about climate change called *An Inconvenient Truth*. The message from the film is clear – the world must change or face devastating consequences.

- Global temperatures have already risen over the past century, and could increase by 0.5°C (9°F).
- Melting glaciers will increase flood risk.
- Crop yields will decrease, particularly in Africa.
- Up to 40 percent of living things could face extinction.
- There will be more extreme weather patterns.

Al Gore insists that action is urgent. Governments need to:

- Promote and invest in cleaner energy and transport technology.
- Substantially reduce carbon dioxide emissions.
- Prevent deforestation.

Climate change – it's not happening!

While most scientists in the world agree that climate change is a real threat caused by human activity, there are still some who do not. Most of these scientists agree that our climate is getting warmer but believe that warming and cooling of our climate happens naturally. We have evidence to support both theories but no absolute proof.

The thawing of the ice around the North and South Poles is evidence of global warming. Those who disagree say that in the 15th century there was practically no ice around the North Pole and that global temperatures vary approximately every 1,000 years.

Too valuable to burn

Get cracking

Some fractions of crude oil are more in demand than others. We want more of the small molecules that fuel our cars and aeroplanes, and less of the big molecule fractions that make up lubricating oil and fuel oil. Very thick oil from some parts of the world does not give us enough of the small molecules we need.

To get around this problem, it is possible to break big molecules into smaller ones. This process is called **cracking**. When a large hydrocarbon molecule is heated with a suitable catalyst, it breaks into smaller molecules. Cracking an alkane produces a smaller alkane and a new type of hydrocarbon called an **alkene**.

 THE SCIENCE YOU LEARN: ALKENES

Alkenes have fewer hydrogen atoms per molecule than alkanes. In these compounds a pair of carbon atoms join together with two bonds. We call these links double bonds. The simplest alkenes are:

Ethene, C_2H_4 Propene, C_3H_6

 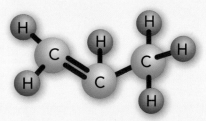

It is possible to make many new substances from alkenes. They are the basis for the petrochemicals industry, which gives us a wide range of useful products.

The discovery of polythene

In England in 1933, two research chemists Eric Fawcett and Reginald Gibson were trying to make interesting new compounds by working at very high pressures. They tried to get ethene to react at 180°C (356°F) under a pressure 2,000 times that of the Earth's atmosphere. Their experiments were disappointing – none of them produced anything useful. However, at the end of one experiment, the walls of the container they were using were coated with a thin layer of a white, waxy solid. The two chemists had created a new material, which they called polythene.

As hard as they tried, they could not repeat the experiment to make more. In fact one attempt led to the violent explosion of their apparatus. In 1935 they tried again with better equipment, but still had no success. Then, during one experiment they had a leak in their equipment and needed to pump in more ethene. This time they made 8 g (0.28 oz) of the white solid. They realised that the reaction only worked if a small amount of oxygen was present. The experiment had been successful because of the leak in the container. This was the beginning of the age of plastics.

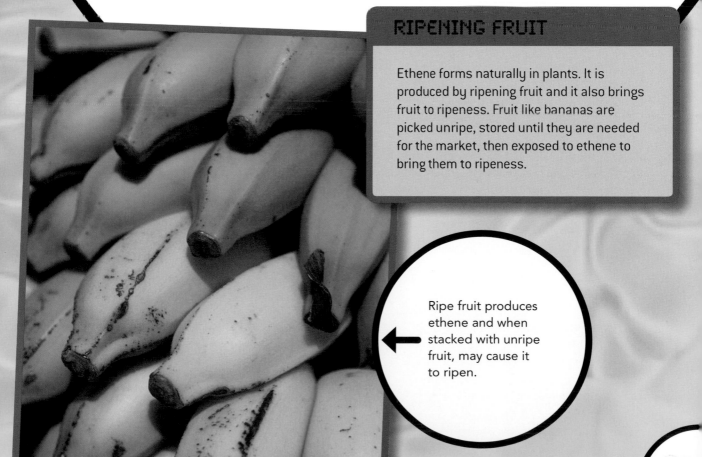

RIPENING FRUIT

Ethene forms naturally in plants. It is produced by ripening fruit and it also brings fruit to ripeness. Fruit like bananas are picked unripe, stored until they are needed for the market, then exposed to ethene to bring them to ripeness.

Ripe fruit produces ethene and when stacked with unripe fruit, may cause it to ripen.

Plastics from oil

Ethene and other alkenes are the starting points to make many plastics. Special conditions – heat, pressure, and a catalyst – are needed to make alkenes **polymerize** (make **polymers** – see box). The boxes on this page show some common polymers made by adding alkene molecules together.

THE SCIENCE YOU LEARN: POLYMERS

"Poly" means many and a polymer is formed when many small molecules join together. The individual molecules are called **monomers**. The simplest, most widely used plastic is polythene (poly-ethene) but there are many more. Alkene molecules polymerize when their double bond breaks and molecules add together in a long chain.

Plastic: Polythene (poly-ethene)
Monomer: Ethene
Uses: Plastic bags, food wrapping

Monomer structure

Polymer chain

Plastic: Polyvinyl chloride (PVC)
Monomer: Vinyl chloride
Uses: Imitation leather, water pipes, window frames, shower curtains

Monomer structure

Polymer chain

Plastic: Polypropene
Monomer: Propene
Uses: Buckets, ropes, textiles, yoghurt pots, straws, bottles and caps.

Monomer structure

Polymer chain

Plastic: Polystyrene
Monomer: Styrene
Uses: Computer cases, drinking cups, moulded parts inside cars. Expanded polystyrene is an excellent insulator.

Monomer structure

Polymer chain

Plastic, plastic everywhere

Plastics are useful because they are cheap and long-lasting. Unfortunately there are two main problems with plastic. Firstly they do not break down easily in the environment and secondly it is difficult to recycle them.

Why don't we recycle more plastics?

The main problem with recycling plastics is that they must be separated into different types and each type processed differently. Some objects are made of two plastics. For example, the cap of a water bottle is different from the bottle itself and this makes separation even more difficult.

Number	Plastic	Items
1 PETE	PET Poly(ethene terephthalate)	Soft drink bottles, water bottles
2 HDPE	HDPE High density poly(ethene)	Containers for detergent and fabric softeners, bleach, and milk
3 V	PVC Poly(vinyl chloride)	Pipes, shower curtains, shrink wrap, seat covers
4	LDPE Low density poly(ethene)	Wrapping bags, grocery bags, sandwich bags
5 PP	PP Poly(propene)	Bowls and buckets
6 PS	PS Poly(styrene)	Disposable cutlery, coffee cups, meat trays, packaging
7	A combination of plastics 1–6 or another less commonly used plastic such as poly(carbonate)	Baby bottles, re-usable sports water bottles

In 1988, plastics manufacturers agreed a labelling system to help people recycle different types of plastic. Look on bottles and packaging materials for the recycling triangle. The number in the centre identifies the type of plastic. Plastic numbers 1, 2, and 3 are the easiest to recycle.

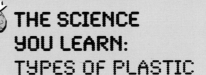

THE SCIENCE YOU LEARN: TYPES OF PLASTIC

There are two main types of plastic. **Thermoplastics** can be heated, softened, and remoulded. Plastics made from alkenes are thermoplastics. **Thermosetting plastics** cannot be melted and reformed. Electrical plugs and sockets are made from thermosetting plastics. Thermoplastics are much easier to recycle than thermosetting plastics.

What happens to waste plastic?

Throw away a polythene bag and it will eventually break down. The trouble is it takes hundreds of years! Authorities in several countries are concerned about manufacturers' claims that plastics are biodegradable. Legal definitions have now been agreed to protect consumers from false claims.

Degradable plastics: These break down into smaller pieces of plastic in the presence of heat and light. They will eventually biodegrade. They are mainly made from polythene and break down in around 18 months.

Biodegradable plastics: These are broken down by the action of microorganisms. However, this can take many years, and in landfill where there is no oxygen, even longer. There is still some debate about the meaning of the term biodegradable plastics.

Compostable plastics: For a plastic to be described as compostable it must biodegrade by 90 percent in 6 months in industrial composting facilities. The plastic eventually decomposes into water, carbon dioxide, and **biomass**. These plastics are not made from oil.

CUTTING EDGE: PHA

PHA stands for polyhydroxyalkanoates. These are biodegradable polymers produced naturally inside bacteria. They are not plastics but have the same qualities. The most common PHA behaves just like polypropene. Scientists believe PHAs could replace the huge amount of oil-based plastics used worldwide.

Could new biodegradable polymers make litter like this a thing of the past?

Degradable polythene

A British company claims to have developed a plastic that degrades totally without affecting the environment. Called d_2w^{TM}, the new plastic has been approved as safe to come into contact with food. The company is already using it to make bread bags, freezer bags, packaging films, and refuse bags.

d_2w^{TM} is produced using degradable, compostable plastics technology or DCP for short. To make the plastic degradable an additive is combined with polythene. The manufacturers claim that the plastic begins to degrade when exposed to heat or light. It eventually breaks down to just carbon dioxide and water.

When manufacturers claim a plastic is biodegradable, what do they actually mean? Fortunately there are now laws that define what is meant by such descriptions.

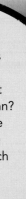

 INVESTIGATION: ROT OR NOT?

Is d_2w^{TM} really as good as the manufacturers claim? How could you find out? There are three things to investigate.

1. Is it really degradable? Plan an investigation to compare how d_2w^{TM} behaves compared to normal plastic bags. How will you carry out a fair comparison? How will you find out what conditions are needed? What about the length of time it takes?
2. Is it really as strong as normal polythene? Devise an investigation to compare the strength of d_2w^{TM} and normal refuse sacks.
3. Finally check its price. If it is too expensive people just will not use it.

What is the future for oil?

Oil is a **non-renewable** fuel and will eventually run out. When it will run out depends upon many factors:

- How much more oil will be found in the Earth's crust?
- How fast will the demand for oil grow?
- How much will prices rise?
- As prices rise, how much of the poor quality oil (for example shale oil) will become economically worthwhile?
- Will advancing technology enable us to extract more oil from existing oil fields?

It is estimated that currently-known oil reserves will last 40 years. If more is found and technology improves, it could easily last 100 years. However, it is likely that production will peak – it might even have peaked already– in the early part of the 21st century. After this oil production will fall, and it will become more expensive.

THE SCIENCE YOU LEARN: SUSTAINABILITY

A definition for sustainability is "providing for the needs of today while caring for the requirements of future generations". Oil use is not sustainable. It will eventually run out and most scientists accept that its use is contributing to climate change.

(Data obtained from the California Energy Commission so relates to the processing of oil in that U.S. state.)

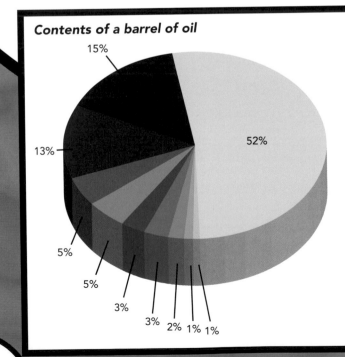

Contents of a barrel of oil

15%
13%
52%
5%
5%
3%
3% 2% 1% 1%

- Finished Motor Gasoline
- Lubricants
- Other Refined Products
- Asphalt and Road Oil
- Liquefied Refinery Gas
- Residual Fuel Oil
- Marketable Coke
- Still Gas
- Jet Fuel
- Distillate Fuel Oil

Alternatives to oil

Many of us use cars, travel to holiday destinations by aeroplane, or buy goods and produce brought from different parts of the world. However, our fossil fuel use is not sustainable. We need to: a) try to reduce our use of fossil fuels, and b) look for alternative ways to power our transport.

CUTTING EDGE: ALTERNATIVE FUELS

Ethanol

Countries with a suitable climate and plenty of space grow sugar cane and **ferment** the sugar to make ethanol (alcohol). Brazil is a major producer of ethanol, which can be blended with petrol to fuel cars.

Biodiesel

This is a renewable fuel that can be manufactured from vegetable oils, animal fats, or recycled restaurant cooking oil. Biodiesel can be used in unmodified diesel engine vehicles.

Biofuels – the downside

Farming for fuels isn't without problems. Already, developing countries are seeing biofuels (fuels grown from crops) as good for their economy and they are keen to grow more. As a result, large areas of rainforest are being cleared to make way for growing palm oil. Corn oil use presents another problem. Food prices have risen in some countries as less corn is being grown for food.

British entrepreneur Sir Richard Branson, seen here with Prime Minister Gordon Brown, launches the first UK train to run on biodiesel.

The perfect answer?

Many motor manufacturers believe the future of fuel lies with hydrogen, which will power cars as effectively as now, but without relying on oil. Most hydrogen is currently produced from fossil fuels but it can be produced from water by a process called **electrolysis** (see box). When hydrogen burns it combines with oxygen to re-make water. It is a perfectly clean fuel.

However, there remain problems. Electrolyzing water requires electricity and most electricity is generated by burning fossil fuels. Technology will have to develop so that solar, wind, and wave energy generate the electricity we need. Hydrogen gas is also harder to store than a liquid fuel. The hydrogen must be compressed (squeezed), which raises safety and cost issues. Once these technological problems are overcome, more cars could run on hydrogen.

Once the problems of generating and storing hydrogen are solved, more of us could drive hydrogen-fuelled cars like this.

THE SCIENCE YOU LEARN:
ELECTROLYSIS OF WATER

Pure water will not conduct electricity but it will if a little acid is added. Dipping two **electrodes** (rods) into slightly acid water and connecting them in a circuit allows a current to flow. The water is chemically changed by the current. This is electrolysis. If a direct current flows through the water, hydrogen is produced at the positive electrode and oxygen at the negative electrode. The electricity splits the water (H_2O) into its elements.

Hydrogen fuel gets shuttles into space. It also drives the fuel cells that power the craft's electrical systems. How long will it be before hydrogen use really takes off for us all?

CUTTING EDGE: A NEW SOURCE OF HYDROGEN

At present, most hydrogen is made from natural gas, and producing it creates carbon dioxide. Researchers at Penn State University, USA, have taken waste water containing starch from the food industry and added hydrogen-producing bacteria. The result is clean water and hydrogen! Similar work at Birmingham University, UK, involved feeding waste from the **confectionery** industry to bacteria. Again, hydrogen was produced and the waste was disposed of cleanly.

A vision for the future

Hydrogen can burn, which means that it is possible to use it in a car's engine. However, there is a more efficient method of obtaining energy from hydrogen – a fuel cell. A hydrogen fuel cell combines oxygen and hydrogen together to make water. There is no flame or explosion. Instead of generating heat, the fuel cell produces electricity, which drives the car's motor.

Technology is developing all the time, the cost of fuel cells has dropped, and most major car manufacturers have hydrogen fuel cell cars in development. The BMW Hydrogen 7 is the first hydrogen-powered luxury car designed for everyday use.

Facts and figures

On these pages you can find out about the careers of some of the people who have had a major influence on the oil industry.

Edwin Drake

In 1859, American engineer Edwin L Drake (1819–80) drilled the first United States oil well near Titusville, Pennsylvania, USA. To do so, he invented the drive pipe, which is an iron pipe made of 3-metre (10-foot) sections. The drive pipe was forced down into the ground. The drilling tools were lowered through the pipe and used to cut through the rock. Drake eventually struck oil at a depth of 21 m (69 ft). Foolishly he did not **patent** his invention. Drake died a poor man.

Karl Freidrich Benz

In 1871, German engineer Karl Benz (1844–1929) founded his first company, which supplied building materials. He then began work on developing an internal combustion engine. He received his first patent in 1879 and founded Benz & Company to produce industrial engines in Germany. He began designing a "motor carriage". Benz designed his three-wheel carriage engine with an electric ignition, differential gears, and water-cooling. It was first driven in Mannheim in 1885. On 29 January 1886, he was granted a patent for his petrol-fuelled automobile. In July of the same year, he began selling his automobile to the public. The motor car had arrived!

Rudolf Diesel

In 1892, German engineer Rudolf Diesel (1858–1913) patented the diesel engine. Unlike the internal combustion engine, this one did not need a spark to ignite the fuel-air mixture. After winning his patent for the diesel engine, he continued to work on its development for years. The diesel engine allowed trains and ships to operate with oil instead of coal. Diesel quickly became a rich man. In 1913, he vanished from a ship crossing the channel to England. His body washed up 10 days later. Some believe he committed suicide, but others believe he was murdered, perhaps by competing industrialists.

Al Gore

American politician Al Gore (born 1948) served as the 45th Vice President of the United States. In 2000, he stood for president but he did not win. Gore has since committed himself to campaigning on environmental issues. In 2006 he starred in *An Inconvenient Truth* discussing global warming and the environment. In 2007, he provided the leadership for the "Live Earth" series of concerts to raise awareness about climate change. Al Gore won the Nobel Peace Prize in 2007.

Stephen Schneider

American scientist Stephen Schneider (born 1945) has advised the staff of six United States presidents on issues relating to climate change. In 1976, his research led him to explain how rising carbon dioxide levels would cause global warming.

Red Adair

The film *Hellfighters* was made in honour of American oil-well firefighter Red Adair (1915–2004). Adair was the son of an immigrant Irish blacksmith. He left school to work as a labourer first on the railways, then in the oil fields. There he learned the necessary techniques before starting his own business as an oil-well fire expert. His successes were spectacular with full details described on http://www.redadair.com/thriller.html.

William D'Arcy

British oil miner William D'Arcy (1849–1917) was born in England. His family emigrated to Australia where he first trained as a lawyer. He then started a gold-mining operation. With a considerable fortune he moved back to England. In 1900, he was persuaded to search for oil in Persia (now known as Iran). After much frustration and several years of searching, he eventually struck oil in 1908. This began the extraction of oil from what is now Iran.

The price of crude oil

During the years after World War II, allowing for **inflation**, the price of oil stayed pretty steady at about U.S. $20-$25 per barrel. Occasionally an international crisis caused a jump in price. The graph below shows how prices have changed and what event has caused a significant leap.

In recent years prices have greatly increased. The reasons for this include increased demand from the growing economies of India and China, a shortage of refineries around the world, and political tensions in the Middle East.

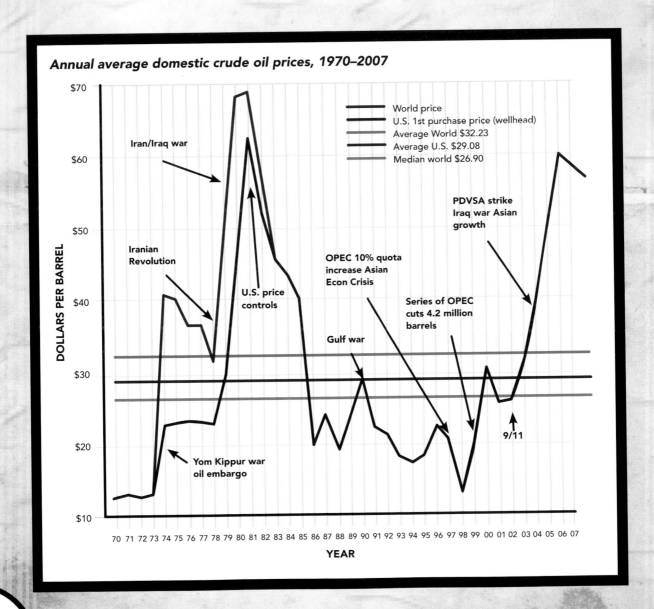

Annual average domestic crude oil prices, 1970–2007

Legend:
- World price
- U.S. 1st purchase price (wellhead)
- Average World $32.23
- Average U.S. $29.08
- Median world $26.90

Labels on graph: Iran/Iraq war, Iranian Revolution, U.S. price controls, Yom Kippur war oil embargo, Gulf war, OPEC 10% quota increase Asian Econ Crisis, Series of OPEC cuts 4.2 million barrels, PDVSA strike Iraq war Asian growth, 9/11

Y-axis: DOLLARS PER BARREL ($10 to $70)
X-axis: YEAR (70 71 72 73 74 75 76 77 78 79 80 81 82 83 84 85 86 87 88 89 90 91 92 93 94 95 96 97 98 99 00 01 02 03 04 05 06 07)

How long will oil last?

The Hubbert peak theory, also known as peak oil, is a theory about the long-term future of oil. It was devised by Dr M King Hubbert. The theory predicts that world oil production will reach a peak and then decline as oil reserves are used up. A lot of controversy surrounds the theory. When the global peak will actually take place depends on past production and on any new discoveries that are made.

The world's largest oil fields

Name	Country	Discovered	Estimated size (billion barrels)	Peak production (million barrels per day)
Gawar	Saudi Arabia	1948	80	5.7
Burgan	Kuwait	1938	70	2.1
Cantarell	Mexico	1976	35	2.2
Bolivar Coastal	Venezuela	1917	32	2.6
Safaniya-Khafji	Saudi Arabia	1951	30	1.5

The octane rating of fuel

When you fill your car with fuel what do the numbers on the pumps mean? They describe the octane number of the fuel. Find out what this is and why is it important.

The alkane series

Alkanes are a series of hydrocarbons. The first four members of the series were listed on page 21. Larger alkanes are:

Pentane	C_5H_{12}	Decane	$C_{10}H_{22}$
Hexane	C_6H_{14}	Pentadecane	$C_{15}H_{32}$
Heptane	C_7H_{16}	Icosane	$C_{20}H_{42}$
Octane	C_8H_{18}	Pentacosane	$C_{25}H_{52}$
Nonane	C_9H_{20}	Triacontane	$C_{30}H_{62}$

Find out more

Books

A Short History of Nearly Everything, Bill Bryson (Doubleday, 2003)
In this wonderful book, Bill Bryson doesn't talk about the oil industry but does consider the geology of the Earth and the power of earthquakes. You'll get some understanding of how the Earth's crust has folded in the past creating the all important oil traps.

The Seven Sisters, Anthony Sampson (Hodder and Stoughton, 1975)
This tells the story of the seven largest oil companies and how they changed the world.

The Prize: The Epic Quest for Oil, Money, and Power, Daniel Yergin (Simon & Shuster, 1991)
A history of the worldwide oil industry from the 1850s to 1990.

Oil Rig Roughneck, Geoffrey M. Horn (Gareth Stevens Publishing, 2008)
Describes the work that oil rig crews do, and explains the structure of the oil industry, life on a rig, offshore drilling, and Middle Eastern oil.

Oil Spill, Christopher F. Lampton (Millbrook Press 1992)
Describes how and why oil spills happen, the damage such accidents can do to the environment, and methods used to clean up spills.

Websites

- www.energyquest.ca.gov/story/index.html
  The Energy Story covers fossil fuels, alternative energy
  sources, energy conservation, and prospects for the future.

- www.petrostrategies.org
  Follow the Learning Center link to find out about the
  oil industry.

- www.wpbschoolhouse.btinternet.co.uk/page04/OilProducts.htm
  Use Dr Brown's Chemistry Clinic to find out all about oil
  and hydrocarbons.

- http://thinkquest/library/search.html
  Search for Exxon Valdez to read *Prince William Sound:
  Paradise Lost?* This explains why the oil spill happened and the
  long-term impact on the environment.

- www.cnergyinst.org.uk/education/glossary
  All the terms used in the oil industry explained simply.

- http://home.utah.edu/~ptt25660/tran.html
  A simple look at alternative sources of energy.

- www.inetec.co.uk/pages/alternative-energy-sources.php
  A detailed look at alternative sources of energy.

- http://fossil.energy.gov/education/energylessons/index.html
  Learn about fossil fuels with the US Dept of Energy.

Glossary

acid rain   rain made acidic due to the presence of sulphur dioxide from coal burning, and nitrogen oxides from car exhausts and other sources

alkanes   compounds made of only carbon and hydrogen with the general formula C_nH_{2n+2}

alkenes   compounds made of only carbon and hydrogen with the general formula C_nH_{2n}. These compounds all have a double bond between two carbon atoms.

anticline   layers of rock folded into a cone shape

biomass   living or recently dead biological material that can be used for energy production or as a raw material for industrial processes

catalyst   substance that speeds up a chemical reaction without being used up itself. At the end of a reaction the catalyst is chemically unchanged.

confectionery   sweets and chocolates

cracking   method for breaking down complex molecules into simpler molecules

crude oil   (also called petroleum) naturally-occurring black or brown liquid found in the Earth's crust

distill   boil a liquid mixture, and then condense the vapour to separate the pure liquid

electrode   rod that takes electricity into a non-metallic part of a circuit

electrolysis   passage of electricity through a liquid that causes a chemical change to occur

exothermic   chemical reaction that releases energy in the form of heat

fault   place where layers of rock in the Earth's crust break and slip past each other

ferment   use microorganisms (such as yeasts or bacteria) to get energy from a biological material in the absence of oxygen

fossil fuel   coal, oil, and gas found naturally in the Earth formed from the remains of dead animals and plants over millions of years

fractional distillation   separation of a mixture of liquids according to their different boiling points

hydrocarbon   chemical compound made of only carbon and hydrogen

impermeable   substance, such as clay, that will not let liquids soak through it

inflation   rise in the general level of prices of goods and services over a period of time

monomer small molecule that will bond wih other monomers to form a long chain

non-renewable resource that cannot be re-made, re-grown, or regenerated on a scale comparative to its consumption

patent set of rights granted to an inventor for a period of time that prevents others from copying the invention

petroleum (also called crude oil) naturally-occurring black or brown liquid found in the Earth's crust

plastic range of synthetic (man-made) substances made by joining many small molecules together to create long-chain molecules

pollutant chemical substance introduced into the environment that has harmful effects on human health, other living organisms, or the environment

polymer substance that is composed of large, long-chain molecules made up from a repeating unit

polymerize bonding many small molecules together to form a long chain

porous rock solid rock that has a network of tiny pores (holes) that can soak up liquid like a sponge

recycle process of turning materials into new products. This prevents the waste of useful substances and energy.

reservoir rock porous rock formations that contain oil

sedimentary rock rock formed by small particles of other rocks, or from particles of once living things, compacted together over millions of years

seismogram graphic record of the movement of the Earth's crust after an earthquake or after artificially-created shock waves

seismology study of vibrations through the Earth's crust. Earthquakes naturally create vibrations but scientists also create vibrations to learn about the layers of rock in the Earth

synthetic something made from an artificial or man-made substance

tar sands sandstone in which oil is trapped

thermoplastic type of plastic that will melt to a liquid when heated, and freeze back to a solid when cooled

thermosetting plastic plastic that cannot be reshaped by heating. It is formed into a particular rigid shape when it is made and this cannot subsequently be changed.

toxic substance that is harmful to health or lethal if it enters the body in large quantities

viscosity the thickness of a liquid